Electronic
abbreviations
symbols
and terms

THORN EMI HOME ELECTRONICS/MACMILLAN SERIES

Books
Electronic abbreviations, symbols and terms
VHS recording principles

An introduction to digital techniques books
An introduction to digital techniques: logic gates, flip-flops, counters and shift registers
Digital displays and applications: decoders and encoders, memories, text data acquisition, viewdata
An introduction to microprocessors: microprocessors, programs, interfacing, timesharing, data transfer

Tutor boards
Digital tutor board with exercise book
Transistor tutor board with exercise book

Video tapes
Introduction to microprocessor systems
Introduction to the teletext system
VHS mechanical principles

For the tutor
Two-day teletext course
Basic VHS course

For everyone
Basic mathematical skills

Electronic abbreviations, symbols and terms

© THORN EMI Television Rentals Ltd 1984, 1985

All rights reserved. No reproduction, copy or transmission of this publication may be made without written permission.

No paragraph of this publication may be reproduced, copied or transmitted save with written permission or in accordance with the provisions of the Copyright Act 1956 (as amended).

Any person who does any unauthorised act in relation to this publication may be liable to criminal prosecution and civil claims for damages.

First published 1984

Published in this format in 1985 by
MACMILLAN EDUCATION LTD
Houndmills, Basingstoke, Hampshire RG21 2XS
and London
Companies and representatives
throughout the world

SECOND EDITION

Printed in Hong Kong

British Library Cataloguing in Publication Data
Electronic abbreviations, symbols and terms. —
(THORN EMI Home Electronics series)
1. Electronics—Abbreviations
I. Series
621.381'0148   TK7804
ISBN 0-333-39167-5

© THORN EMI Television Rentals Ltd 1984, 1985

All rights reserved. No reproduction, copy or transmission
of this publication may be made without written permission.

No paragraph of this publication may be reproduced, copied
or transmitted save with written permission or in accordance
with the provisions of the Copyright Act 1956 (as amended).

Any person who does any unauthorised act in relation to
this publication may be liable to criminal prosecution and
civil claims for damages.

First published 1984

Published in this format in 1985 by
MACMILLAN EDUCATION LTD
Houndmills, Basingstoke, Hampshire RG21 2XS
and London
Companies and representatives
throughout the world

Printed in Hong Kong

British Library Cataloguing in Publication Data
Electronic abbreviations, symbols and terms.—
(THORN EMI Home Electronics series)
1. Electronics—Abbreviations
I. Series
621.381'0148   TK 7804
ISBN 0-333-39167-5

CONTENTS

Introduction 7
Abbreviations (alphabetical order) 2
Symbols commonly used Greek letters 9
mathematical 9
CRTs 10
capacitors 12
coils, transformers 14
fuses, resistors 16
waveforms 19
battery 21
semiconductors 22
miscellaneous 31
filters 35
logic 35
data processing flowchart 39
cable distribution systems 40
Terms video recording 48
microprocessing 57
Appendix component colour code system 67
Index 74

INTRODUCTION

Listening to electronics technicians discussing their work must, to the layman, be like listening to a foreign language; a circuit diagram must appear to him like a strange sort of street map. In the case of the written word, even more mystery can abound because words are often abbreviated, sometimes to just one letter.

It would become unwieldy to use some technical words every time they appear in a circuit diagram or elsewhere, so such abbreviations are essential. Circuit symbols represent components to the electronics technician in a way that a drawing of the device could not.

So that technicians can communicate effectively it is essential that the same abbreviations, symbols and terms are consistently used.

Most technical terms, abbreviations and symbols used in electronics are the subject of agreed standards set by the British Standards Institute and where these exist they should be used. However, in the case of logic symbols the electronics industry in Britain appears to have adopted the American symbols whilst technical colleges teach and use British Standards.

This booklet contains many of the abbreviations, symbols and terms that an electronics technician or student is likely to come across.

ABBREVIATION	UNIT/MEANING
A	ampere (unit of current)
A	anode (valve, diode etc)
A	alpha (Greek alphabet)
ac	alternating current
acc	automatic colour control
Ae	aerial
af	audio frequency
afc	automatic frequency control
agc	automatic gain control
alc	automatic level control
am	amplitude modulation
apc	automatic phase control
ASCII	American Standard Code for Information Interchange
avc	automatic volume control
B	beta (Greek alphabet)
B	flux density
b	base of a transistor
bcd	binary coded decimal
bfo	beat frequency oscillator
bpf	band pass filter
brm	bit rate multiplier
BST	British Standard Time
BTU	British Thermal Unit
C	capacitor
C	coulomb (unit of charge)
c	collector (of a transistor)

ABBREVIATIONS

CCTV	Closed Circuit Television
cd	candela (luminous intensity)
CMOS	Complementary Metal Oxide Semi-conductor
cps	cycles per second
CPU	Central Processing Unit
CRT	Cathode Ray Tube
cs	cycles per second
cw	continuous wave
cw	carrier wave
d	drain (of an fet)
dB	decibel
dc	direct current
dl	delay line
DIL	Dual In Line
DPDT	Double-Pole Double Throw
DPST	Double-Pole Single Throw
E	electromotive force
e	emitter
ef	emitter follower
EHT	Extra High Tension
erp	effective radiated power
f	farad
f	frequency
fet	field effect transistor
ff	flip-flop
fg	frequency gearing

fm	frequency modulation
fo	resonant frequency
fT	frequency cut-off point (of a transistor)
G	Giga (used to denote 10^9 or 1 000 000 000)
G	Grid (of a valve)
g	gate (of SCR or fet)
GMT	Greenwich Mean Time
H	Henry
H	horizontal polarization
H	magnetic field strength
hf	high frequency
hfe	transistor small signal current gain
hFE	transistor large signal current gain
hpf	high pass filter
HT	High Tension
Hz	Hertz
I	current
Ib	transistor base current
IC	Integrated Circuit
IF	Intermediate Frequency
Ic	transistor collector current
Ie	transistor emitter current
I/P	input
ips	inches per second
J	Joule

ABBREVIATIONS

K	Kelvin
k	kilo (used to denote 10^3 or 1000)
k	cathode (valve, diode etc)
kHz	kilohertz
kW	kilowatt
L	inductor
LCD	Liquid Crystal Display
LED	Light Emitting Diode
lf	low frequency
lm	lumen
lpf	low pass filter
LS	loudspeaker
LSI	Large Scale Integration
LW	Longwave
M	Mega (used to denote 10^6 or 1 000 000)
m	metre
m	milli (used to denote 10^{-3} or $\frac{1}{1000}$)
mA	milliamp
mf	medium frequency
mfd	microfarad
mH	millihenry
MHz	Megahertz
mm	millimetre
mpu	microprocessor unit
ms	millisecond
mV	millivolt
mw	milliwave

ABBREVIATIONS

mW milliwatt
MW megawatt

N Newton (force)
N Negative, Neutral
n nano (used to denote 10^{-9} or $\frac{1}{1\,000\,000\,000}$)
nf nanofarads
NMOS 'N' type Metal Oxide Semiconductor
npn type of transistor
ns nanosecond
NTSC National Television System Committee

o/c open circuit
ohm unit of resistance
O/P output

P Positive
P Power
p pico (used to denote 10^{-12} or $\frac{1}{1\,000\,000\,000\,000}$)
PA Public Address
PAL Phase Alternate Line
PB Playback
pd potential difference
pf picofarad
PISO Parallel In Serial Out
pnp type of transistor
Po output power
PU Pick-Up

Q magnification factor
Q quantity of electricity (in coulombs)

ABBREVIATIONS

R	Resistor
RAM	Random Access Memory
rf	radio frequency
rfc	radio frequency choke
RL	load resistance
rms	root-mean-square
ROM	Read Only Memory
Rx	receiver

s	second
s	source (of an fet)
s/c	short circuit
SCR	Silicon Controlled Rectifier
SECAM	Séquential Couleur A Mémoire
SIPO	Serial In Parallel Out
SPDT	Single Pole Double Throw
SPST	Single Pole Single Throw
sw	short wave
swg	standard wire gauge

T	Tera (used to denote 10^{12} or 1 000 000 000 000)
T	Teala (magnetic flux density)
t	time
tr	rise time
trf	tuned radio frequency
Tx	transmitter

UHF	Ultra High Frequency

ABBREVIATIONS

R	Resistor	V	Volts
RAM	Random Access Memory	V	Vertical polarization
rf	radio frequency	VA	Volt Amps
rfc	radio frequency choke	vhf	very high frequency
RL	load resistance	vlf	very low frequency
rms	root-mean-square		
ROM	Read Only Memory	W	Watt
Rx	receiver	Wb	weber
		X	reactance
s	second	X	axis line
s	source (of an fet)	XC	capacitive reactance
s/c	short circuit	XL	inductive reactance
SCR	Silicon Controlled Rectifier	Xtal	crystal
SECAM	Sequential Couleur A Mémoire		
SIPO	Serial In Parallel Out	Y	axis line
SPDT	Single Pole Double Throw	Y	luminance signal
SPST	Single Pole Single Throw		
sw	short wave	Z	impedance
swg	standard wire gauge	Z	non linear resistance

T	Tera (used to denote 10^{12} or 1 000 000 000 000)
T	Tesla (magnetic flux density)
t	time
tr	rise time
trf	tuned radio frequency
Tx	transmitter

UHF	Ultra High Frequency

SYMBOLS — commonly used Greek letters, mathematical

α	alpha (current amplification factor common base amplifier)
β	beta (current amplification factor common emitter amplifier)
λ	lambda (wavelength)
μ	mu (micro 10^{-6})
μf	microfarad
μV	microvolts
μA	microamps
μH	microhenry
π	pi
Ω	omega, ohm
$\neq$	is not equal to
$=$	equal to
$\approx$	approximately equal to
$>$	greater than
$<$	less than
$\sim$	alternating current
$\simeq$	alternating current and direct current
$\leqslant$	is less than or equal to
$\geqslant$	is greater than or equal to
∞	infinity
$\times$	multiply
$+$	add
$-$	subtract (minus)
$\sqrt{}$	square root

SYMBOLS – CRTs

Character display tube

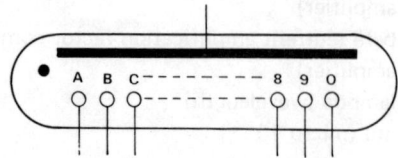

Internal conductive coating

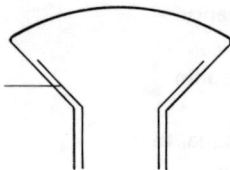

Indirectly heated cathode with associated heater

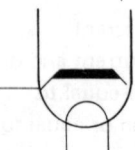

Control grid

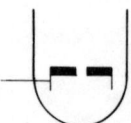

SYMBOLS – CRTs

Focusing electrode

Electron gun (simplified)

Coils for electromagnetic deflection

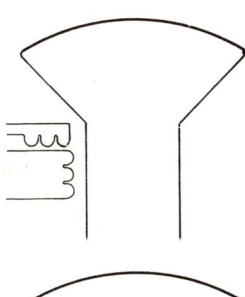

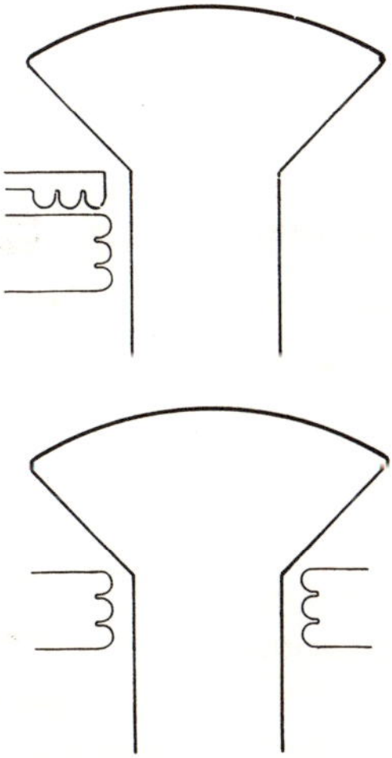

SYMBOLS — capacitors

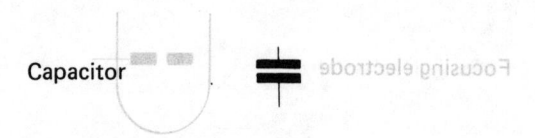

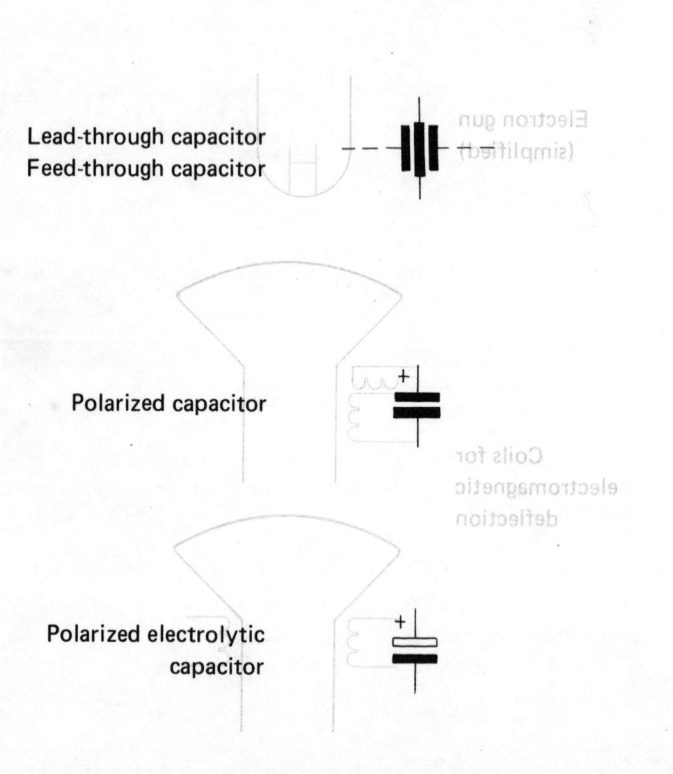

Capacitor	━━ ━━
Lead-through capacitor Feed-through capacitor	▬┃▬
Polarized capacitor	+ ▬▬
Polarized electrolytic capacitor	+ ▭ ▬

SYMBOLS — capacitors

Non-polarized electrolytic capacitor

Variable capacitor

Capacitor with preset adjustment

Temperature-dependent capacitor

14   SYMBOLS — coils, transformers

Winding (inductor, etc.)

Winding with tappings

Winding with core

Transformer

SYMBOLS — coils, transformers

Inductor with variable inductance

Inductor with preset adjustments

Saturable inductor

SYMBOLS — fuses, resistors

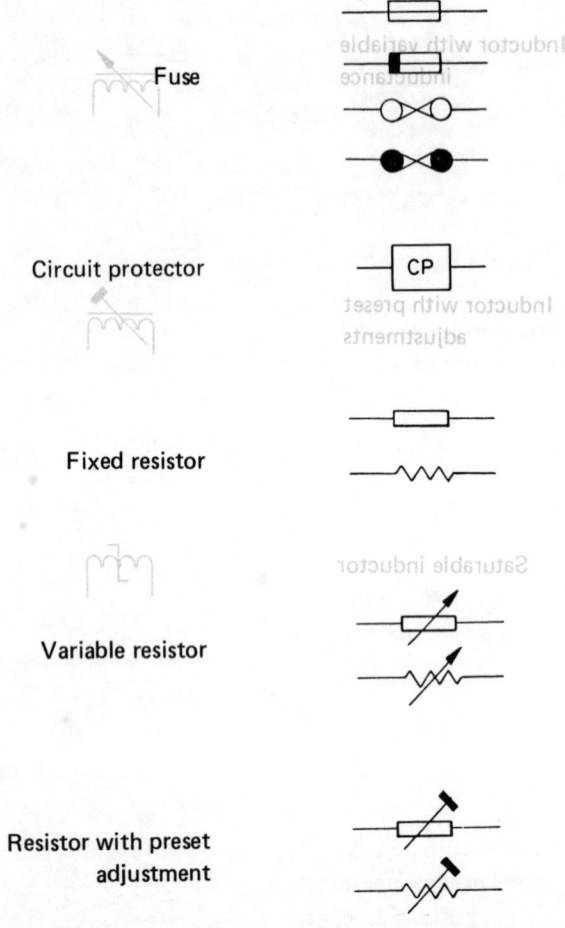

SYMBOLS – resistors

Resistor with moving contact

Resistor with pronounced positive temperature coefficient

Voltage divider with moving contact

Resistor with pronounced negative temperature coefficient

Light sensitive resistor

Voltage divider with preset adjustment

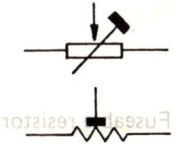

Fusable resistor

Resistor with inherent non-linear variability

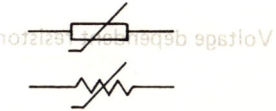

Voltage dependent resistor

SYMBOLS – resistors

Resistor with pronounced positive temperature coefficient

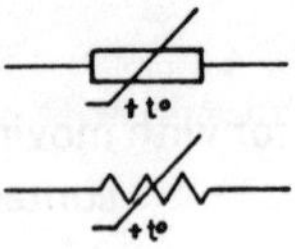

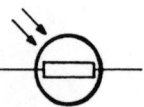

Resistor with pronounced negative temperature coefficient

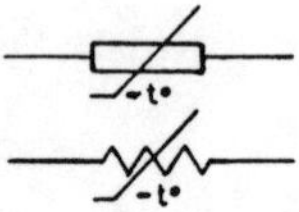

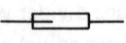

Light sensitive resistor

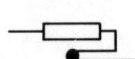

Fuseable resistor

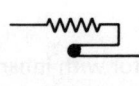

Voltage dependent resistor

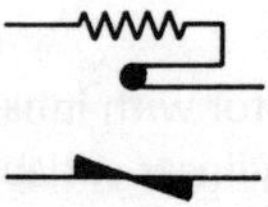

SYMBOLS — waveforms

Pulse positive going

Negative going

Pulse alternating current

Sawtooth

Step positive going

Step negative going

SYMBOLS — waveforms

Pulse position or pulse phase modulation	Pulse positive going
Pulse frequency modulation	Negative going
Pulse amplitude modulation	Pulse alternating current
Pulse duration modulation	Sawtooth
Pulse interval modulation	Step positive going
Pulse code modulation	Step negative going

SYMBOLS — battery   21

Primary or secondary cell

Battery or primary or secondary cells

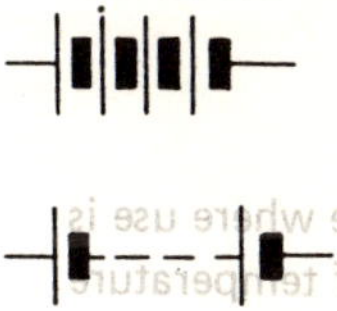

Diode (may be used without the circle)

Diode where use is made of temperature dependence characteristic

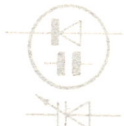

Diode used as a capacitive device (varicap diode)

Tunnel diode

Zener diode

SYMBOLS — semiconductors

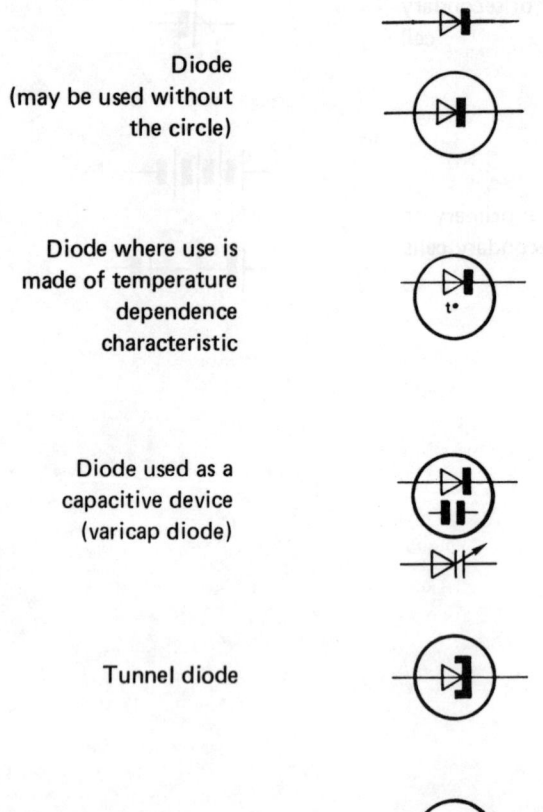

Diode
(may be used without the circle)

Diode where use is made of temperature dependence characteristic

Diode used as a capacitive device (varicap diode)

Tunnel diode

Zener diode

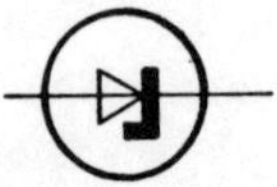

SYMBOLS — semiconductors

Clipper diode

Backward diode

Bi-directional diode

Thyristor

Reverse-blocking diode thyristor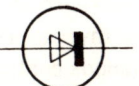

SYMBOLS — semiconductors

Reverse conducting diode thyristor

Bi-directional diode thyristor

Reverse-blocking triode thyristor N-gate (anode controlled)

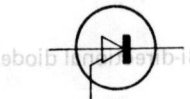

Reverse-blocking triode thyristor P-gate (cathode controlled)

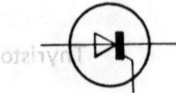

Reverse-conducting triode thyristor N-gate (anode controlled)

SYMBOLS — semiconductors

Reverse-conducting triode thyristor P-gate (anode controlled)

Turn-off triode thyristor N-gate (anode controlled)

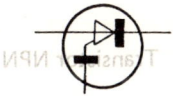

Turn-off triode thyristor P-gate (cathode controlled)

Reverse-blocking thyristor tetrode (SCS)

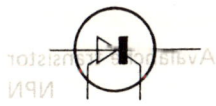

Bi-directional triode thyristor

SYMBOLS — semiconductors

Transistor PNP

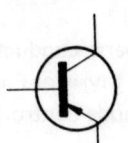

Transistor NPN

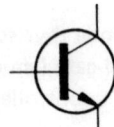

Transistor NPN with collector connected to envelope

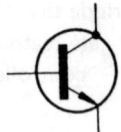

Avalanche transistor NPN

Unijunction transistor P-type base

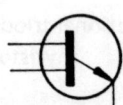

SYMBOLS — semiconductors

Unijunction transistor
N-type base

Transistor with
transverse biased base
NPN

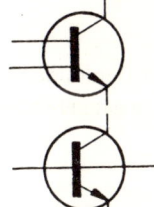

Transistor with ohmic
connection to the I
region PNIP

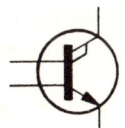

Transistor with ohmic
connection to the I
region PNIN

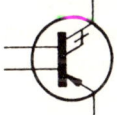

JUG Fet N-channel

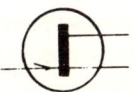

SYMBOLS — semiconductors

JUG Fet P-channel

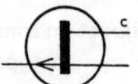

IG Fet Depletion type.
Single gate N-channel

IG Fet Depletion type.
P-channel

IG Fet Depletion type
N-channel Substrate
internally connected

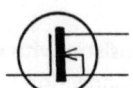

IG Fet Depletion type
P-channel Substrate
connection brought out

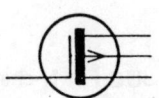

SYMBOLS — semiconductors

IG Fet Depletion type
two-gate N-channel

IG Fet Enhancement type
single-gate P-channel

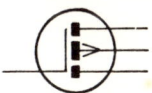

IG Fet Enhancement type
two-gate N-channel

PNP Photo device

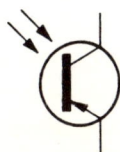

SYMBOLS — semiconductors

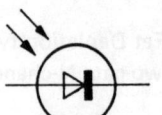

Light-sensitive diode

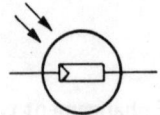

Light-emitting diode

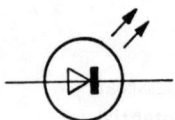

Photo voltage cell

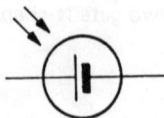

SYMBOLS — miscellaneous

Aerial

Earth

Chassis

Chassis at offset potential

Plug (male)

Plug (female)

Ammeter

SYMBOLS – miscellaneous

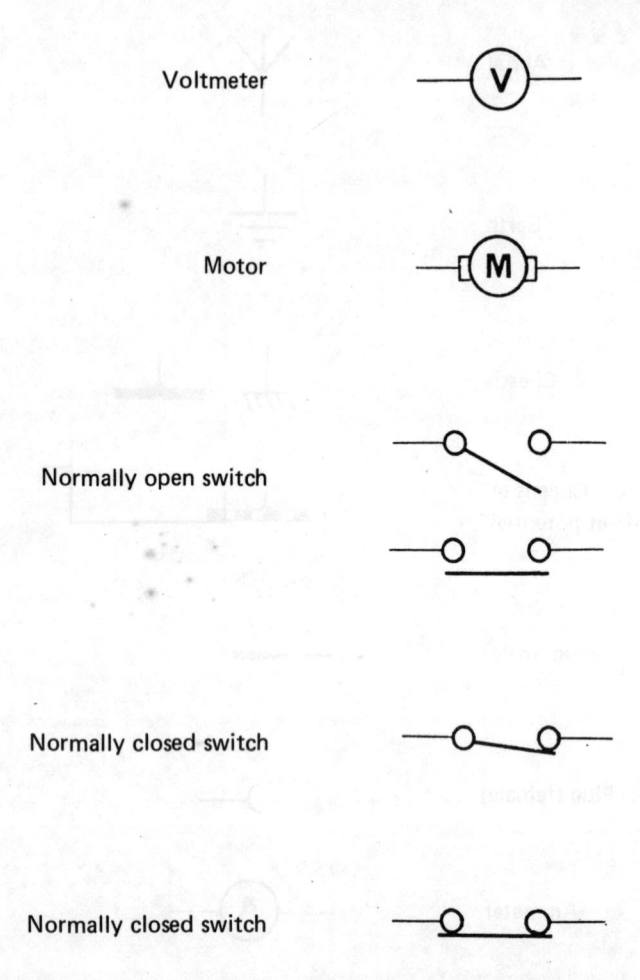

Voltmeter

Motor

Normally open switch

Normally closed switch

Normally closed switch

SYMBOLS — miscellaneous

Relay contact

Electronically operated switch

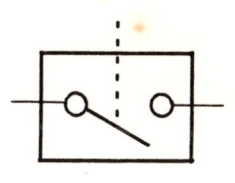

Relay

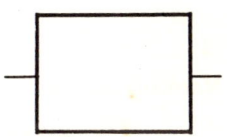

Microphone

SYMBOLS – miscellaneous

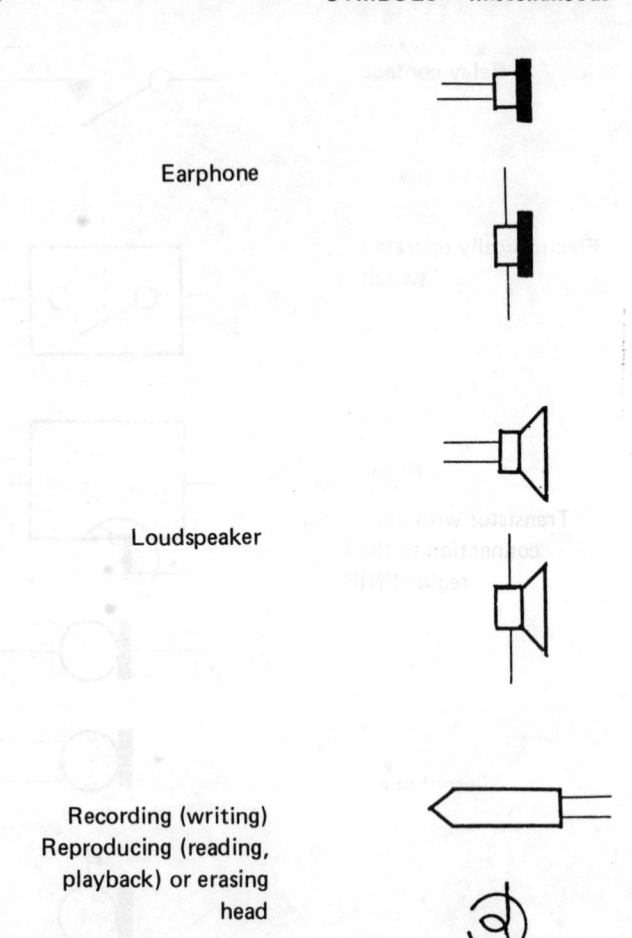

Earphone

Loudspeaker

Recording (writing)
Reproducing (reading,
playback) or erasing
head

SYMBOLS — logic

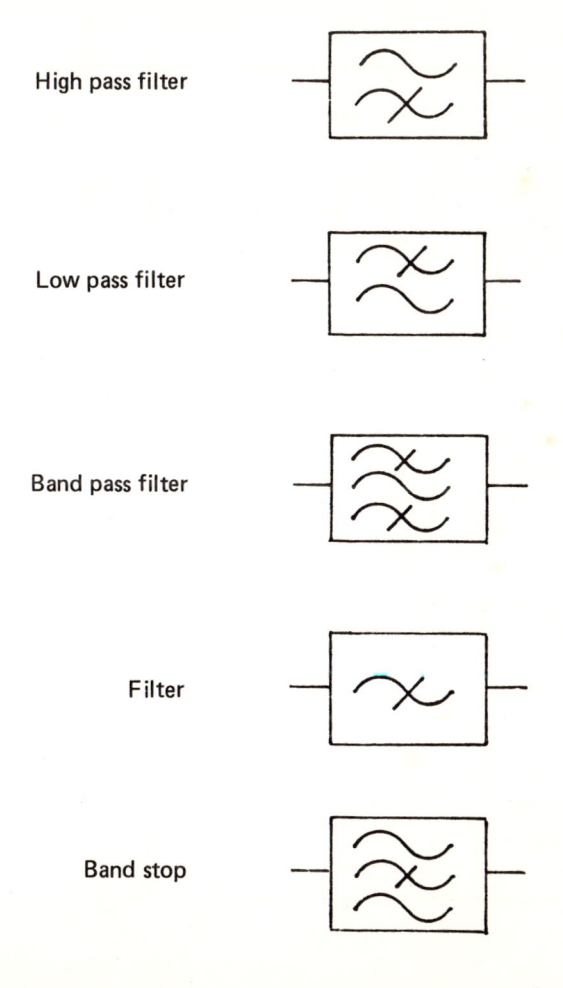

High pass filter

Low pass filter

Band pass filter

Filter

Band stop

SYMBOLS – logic

Gate type	British standard	American	Other symbols
OR	≥1		1
AND	&		● / 2 / &
BUFFER	=		
NOT (INVERTOR)	1		

SYMBOLS – logic

Gate type	British standard	American	Other symbols
NOR	≥1 (with inverter bubble)	(OR shape with bubble)	⊤̄1 / 1
NAND	& (with inverter bubble)	(AND shape with bubble)	&̄ / &
EXCLUSIVE 'OR'		(XOR shape)	e / ⊕

JK flip-flop

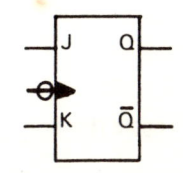

SYMBOLS – logic

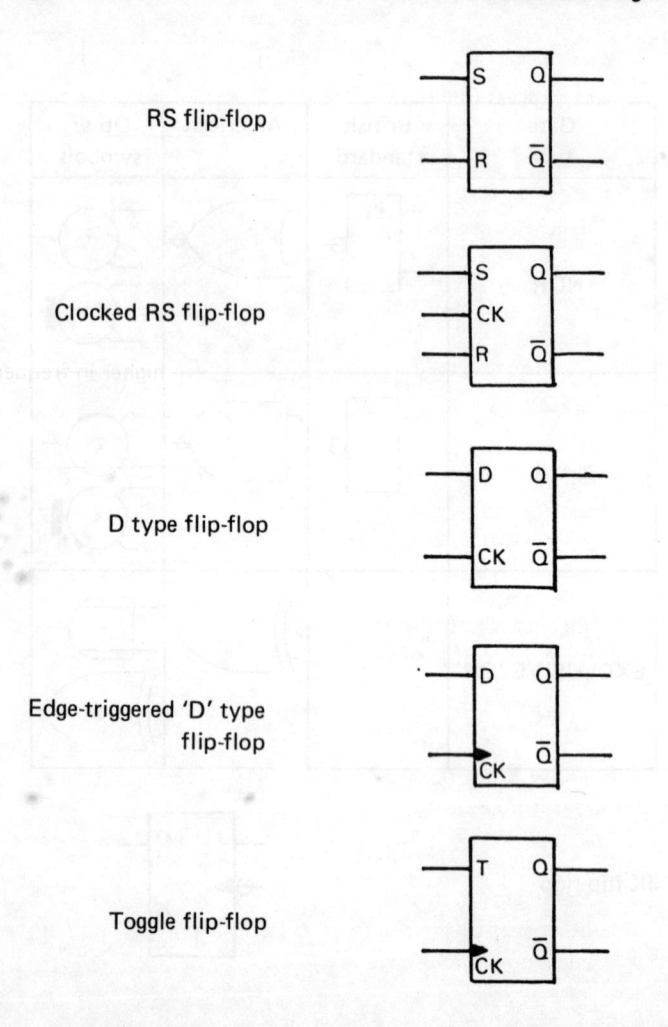

RS flip-flop

S	Q
R	$\bar{Q}$

Clocked RS flip-flop

S	Q
CK	
R	$\bar{Q}$

D type flip-flop

D	Q
CK	$\bar{Q}$

Edge-triggered 'D' type flip-flop

D	Q
CK	$\bar{Q}$

Toggle flip-flop

T	Q
CK	$\bar{Q}$

SYMBOLS — data processing flowchart 39

Any process function

Decision operation

Storage media

Input/output function

Connector to or from another part of the flow chart

Terminator, showing entry to or exit from a procedure

Connector to or from another part of the flow chart on another page

SYMBOLS — cable distribution systems

Receiving aerial

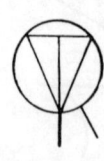

Head ends

Head end employing receiving aerials. (Actual location at the centre of the circle). Trunk output(s) shown at apex of triangle. Distribution outputs at convenient angles to circle only.

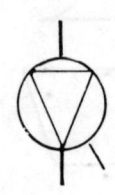

Head end where inputs are not provided by local aerials, (e.g. video feeder or remote aerial site). Input(s) to circle, facing base of triangle, outputs as for above.

Feeder routes

Overhead

Underground

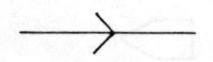

Direction of signal flow

SYMBOLS — cable distribution systems

Amplifiers

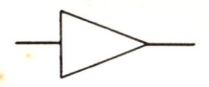

Amplifier with fixed or manually controlled gain and/or gain/frequency characteristic

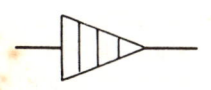

Amplifier with gain and/or gain/frequency characteristic controlled automatically by reference to an external parameter

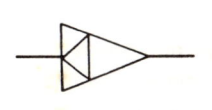

Amplifier with a return channel. Note: (Applicable to all amplifier symbols). Input feeder normal to, and bisecting base of triangle. Output feeder, trunk, to apex of triangle. Output feeder(s) distribution, to side(s) of triangle. (See examples below).

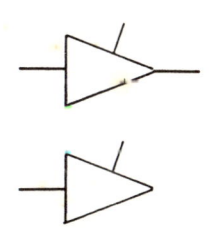

Examples
Trunk/Bridge amplifier, incorporating both trunk and distribution amplifiers

Distribution (branch or spur) amplifier

Bridging amplifier

SYMBOLS – cable distribution systems

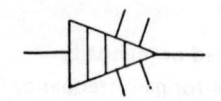

Trun amplifier with automatic level control (ALC) and four distribution outputs

Splitters

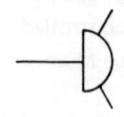

Splitter with equal attenuation to all outputs. Input to straight side, outputs to curved side. Attenuation (dB) to be shown within or adjacent to the symbol (see example)

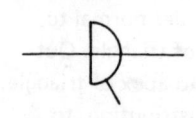

Splitter with one output of minimum attenuation and one or more of higher attenuation, the loss (dB) of which is to be shown within or adjacent to the symbol (see example below)

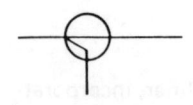

Directional coupler. Side loss in dB to be shown within or adjacent to the symbol

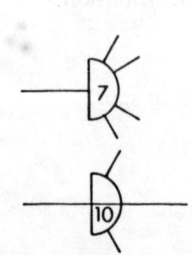

Examples:

Equal four way splitter of 7dB loss to each output

Unequal three way splitter providing two lines at −10dB and a through line

SYMBOLS — cable distribution systems

Subscribers tap-off

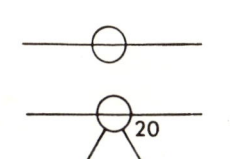

General symbol — single tap-off shown, others may be added as the example below. Attenuation (dB) to tap-off to be shown adjacent to the symbol (see example)

Example — Two subscribers' tap-off at 20dB attenuation

System outlet

Plain (unattenuated) type

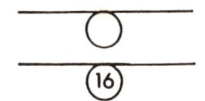

Loop-wired type (including subscriber tap-off.) Attenuation (dB) to be shown within or adjacent to symbol

Equalizers

Attenuation Equalizer

Cable attenuation equalizer for 6dB of cable

Cable simulator (note reversed slope) of 3dB (variable type)

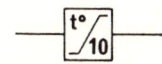

Cable attenuation equalizer for 10dB of cable with thermal compensation

SYMBOLS — cable distribution systems

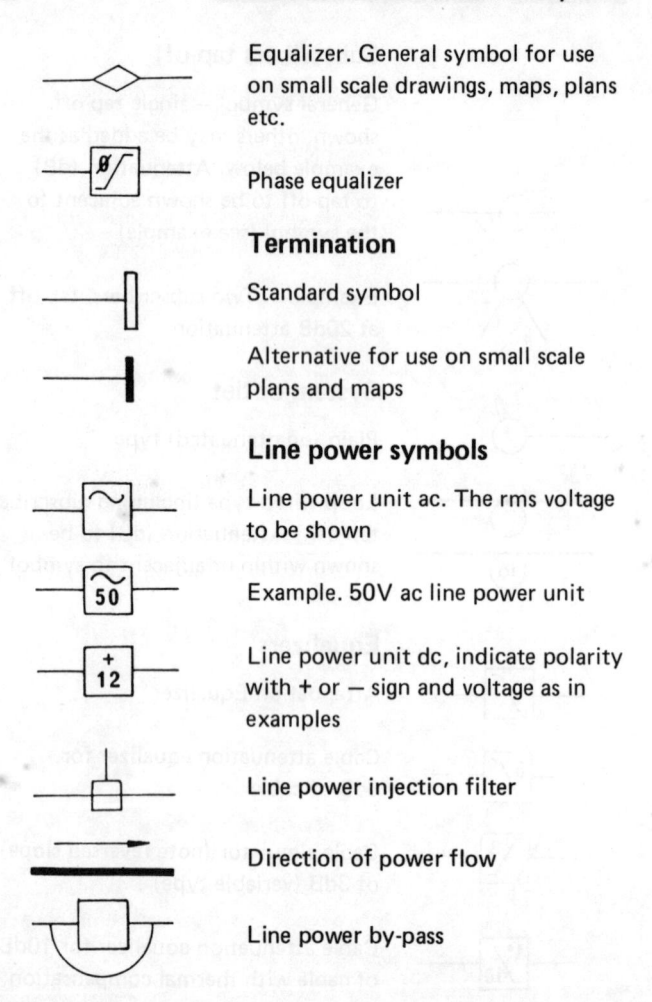

Equalizer. General symbol for use on small scale drawings, maps, plans etc.

Phase equalizer

Termination

Standard symbol

Alternative for use on small scale plans and maps

Line power symbols

Line power unit ac. The rms voltage to be shown

Example. 50V ac line power unit

Line power unit dc, indicate polarity with + or − sign and voltage as in examples

Line power injection filter

Direction of power flow

Line power by-pass

SYMBOLS – cable distribution systems

Line power blocking filter

Examples:

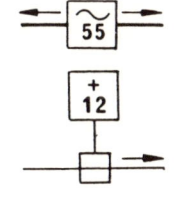

Line power supply unit (55V ac) with integral line power injection feeding two ways

Line power supply unit (12V dc positive) with separate line power injection filter feeding one way only

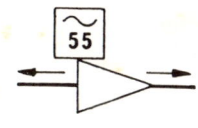

Line power supply unit (55V ac) mounted adjacent to an amplifier and employing line power injection filters integral to the amplifier

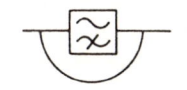

High pass filter with line power by-pass

Radio frequency generators

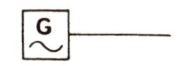

General symbol. Output frequency(ies) to be shown alongside output line(s)

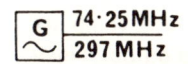

Example. Radio frequency generator for two pilot carriers (combined output) at 74.25 and 297 MHz

SYMBOLS – cable distribution systems

Filters

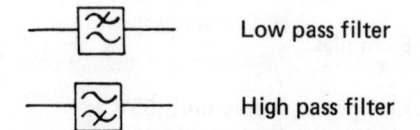

Low pass filter

High pass filter

Band pass filter

Band stop filter

Frequency converter

General symbol, indicate as $f^1 f^2$ the input and output frequencies/channels/bands respectively. If the direction of conversion is not obvious add an arrow to the bottom line as in the example below

Example. Channel converter, input channel 64 to output channel 21

Selective combiner or separator

General symbol, further combined inputs (or separated outputs) may be added as required

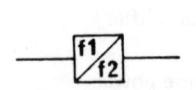

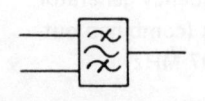

SYMBOLS — cable distribution systems

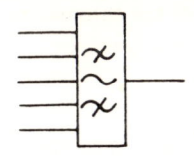

Multiplexer, five-way shown

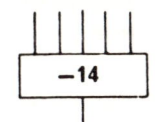

On-selective combiner or separator

Passive types; numerical value of attenuation (dB) to be included as shown

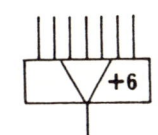

Active types; add amplifier symbol in the appropriate direction and include numerical value of overall gain (dB) as shown

Attenuator

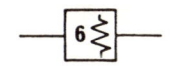

Numerical value in dB to be included within the symbol as shown

TERMS — video recording

ACC. Automatic gain control (using burst) of the chroma signal only, to compensate for different recording sources, e.g. off-air, camera, etc.

AGC. Automatic gain control of the video signal to compensate for different sources, e.g. off-air, camera, etc.

ALC. Automatic Level Control. Dynamic compression of the input audio signal to ensure correct recording level.

Antistatic brush. Earthed brush connecting to slip ring on capstan and drum shafts to ensure perfect earthing at all times.

Audio head. Single record/playback head for linear recording of the audio signal.

Audio track. A portion of the video tape used to record the audio signal. In VHS 1mm wide at the top of the tape. At present linearly recorded.

Auxiliary input. Video or audio input for recording purposes, other than the inbuilt tuner.

Azimuth. The relative angle between the head gap and the signal track is normally 90°.

Back tension. Mechanical 'drag' applied to supply spool during play or fast forward to keep tape taut, and maintain

TERMS — video recording

correct head/tape contact. Achieved on VHS by self-regulating brake band across supply spool.

Belt. Normally rubber, either flat or square in VHS systems.

Betamax. A video tape format engineered by Sanyo, Sony, Toshiba etc. Not compatible with VHS.

Bias signal. Necessary when recording an analogue signal. Consists of an AC signal of constant amplitude many times higher in frequency than the signal to be recorded and is added to it.

Brake. Mechanical – normally spring-loaded felt pad pressed against rotating surfaces, or metal against rubber tyre. Electronic – motor braking using back emf.

Capstan. Polished steel shaft and belt driven by capstan motor which, via the pinch roller, drives the tape.

Chroma under. The conversion of chroma carrier from 4.43MHz to 627kHz (VHS) so that chroma may be recorded without interference with luminance.

Clutch. Slipping drive system (felt against metal) for take-up spool drive, F/F and R/W drive, unload drive. May be adjustable.

Compatibility. The ability to exchange tapes freely between machines of the same make with perfect playback tracking.

TERMS – video recording

Control head. To record and pick up control pulses. Normally physically part of the audio head assembly.

Control pulses/Control track. 25Hz pulses linearly recorded across the bottom of the tape to lock drum or capstan servo (varies with model) on playback.

Counter. Mechanical – belt driven from take up spool with mechanical reset to indicate tape usage/recording start position.

Crosstalk. Pick-up of unwanted signal from adjacent video track due to mechanical/servo limitations.

Dark clip. Removal of overshoots (caused by pre-emphasis) which fall below sync tip and would over-drive the FM modulator (see also White clip).

De-emphasis. Playback process (reduction in HF video amplitude) to counteract the pre-emphasis used on record (See **pre-emphasis**).

Delay equaliser. Playback circuit designed to counteract the different signal delays dependent upon frequency and bandwidth.

Delay line demodulator. Special type of FM demodulator used in certain VHS machines.

TERMS — video recording

Drop out. Momentary loss of video on playback due to oxide missing from tape, loss of head/tape contact, etc.

Drop out compensator. Electronic substitution of signal 'drop out' by a 64μS storage system.

Drum. Rotating polished metal cylinder containing two video heads 180° apart around which tape is wrapped during record and playback. May be belt-driven or directly driven by drum motor.

Dubbing. The facility to replace the audio tracks of a recorded programme with another.

E-E. Electronic-electronic signal. The signal selected by the tuner of the video recorder as viewed on a monitor TV.

End sensors. System of detecting start and end of tape. Usually optical.

Equalisation. Correction by filters or selective gain control to achieve flat record or playback frequency response.

Erase. The removal of magnetic information prior to re-recording. May be full erase (whole tape width) or audio erase (audio track only).

Erase head (see **Erase**). Head may be greater than tape width (full erase) or narrower (1mm audio erase).

TERMS — video recording

Fast forward. Rapid tape transport – the same direction as play, which transfers tape from supply to take-up spool.

Fast forward idler. A clutch-driven rotating wheel which may be moved against the take-up spool to provide fast-forward motion.

Flywheel. Large circular wheel, usually belt-driven, used to help provide constant tape transport or drum speed. (See **Inertia**.)

Freeze frame. Repetitive playback of one complete *picture* (i.e. two consecutive fields).

Guard band. A system to prevent cross-tracking where un-recorded sections were left between adjacent tracks. No longer used in domestic equipment due to uneconomic use of tape.

Guide. Fixed guide — straight polished steel pin; not adjustable. Full guide (or height guide). Shaped guide – (sometimes with built-in oxide trap) to control height of tape; adjustable. Roller guide – nylon roller on metal 'height guides'; adjustable.

H. Commonly used abbreviation for '1 horizontal time period' or '64μs'. Also 'H' may be seen in textbooks as symbol for magnetising force.

Head-to-tape speed. The *relative* speed between the tape and the video heads on record or playback.

TERMS — video recording

Helical scan. System used to achieve high/tape speed required for video. Due to the geometry of tape/head drum, information is recorded as a series of 'diagonal' tracks.

Helical wrap. The system of wrapping the tape around the drum, in the form of a partial helix.

Helical wrap error. If the tape height entering or leaving the drum is incorrect, the FM playback signal will not be of constant amplitude during a field period.

High energy tape. Due to the material used (cobalt doped ferric oxide), achieves a higher retention of magnetic information than normal tape.

Impedance roller. Large circular rollers of metal or plastic resting against the tape and driven by them. Because of their inertia they tend to remove wow and flutter from the tape path.

Inertia. The inability rapidly to change the velocity (in the case of video recorders) of heavy rotational masses. Used to help provide constant tape speed.

Motor discriminator amplifier. An electronic regulator providing coarse correction of capstan and drum motor speed. Fine speed (or phase) control is achieved with a phase-locked loop.

Pause. The stopping of linear tape transport to enable either simple editing of unwanted material on record, or in certain models, a still picture in playback. Operation may be mechanical or electro-mechanical.

Pause solenoid. A solenoid which causes the pinch roller to contact the capstan shaft and therefore drive the tape. Pressing the pause button causes the solenoid to be de-energised.

Phase shift colour recording. An electronic system applied to the chroma signal to prevent the effect of chroma cross-tracking. (Similar to the PAL phase error cancellation system but over a two-line period).

Pre-emphasis. The boosting of high frequency video components prior to recording in order to improve the signal-to-noise ratio on playback (see de-emphasis).

Q. The damping factor applied to the resonant head/transformer assembly on playback. Normally adjustable and may compensate to a degree for normal head wear.

Rewind. Anti-clockwise rotation of the supply spool via an idler pulley will cause tape to transfer rapidly from take-up to supply spool.

Rotary transformer. Method used to couple the signals to and from the video heads.

Servo. Electro-mechanical phase locked loop to produce an exact rotational speed or angular displacement.

TERMS — video recording

Skew error. Bending of vertical picture lines particularly at the point of head switching on the extreme top of the picture.

Slant azimuth. The name given to the system of +6° tilt on the video heads to cancel luminance cross-tracking.

Stop solenoid. A solenoid fed from the system control circuitry which will mechanically cancel any key causing the 'stop' mode.

Tape wrap. See helical wrap.

Tension. See back tension.

Timebase error. If the video information is not played back at a constant rate then 'frequency modulation' of the sync pulses will occur causing sync problems on the monitor.

Tracking. Causing the video heads to read the tape pattern precisely as recorded by correction applied to either drum or capstan servo. A customer tracking control permits the playback of tapes made on other machines.

VHS. The format name.

Video heads. Two ferrite heads mounted on opposite sides of the head drum, which record and playback the video (luminance and chrominance) signal.

White clip. Removal of transients (caused by pre-emphasis) greater than peak white level, which would over-modulate the FM modulator in record. (See **Dark clip**).

Wow and flutter. Tape speed variations causing a change in the frequency of playback of a constant test tone.

Writing speed. The head-tape speed during record.

TERMS — microprocessor

Accumulator. General purpose register used in conjunction with the ALU. When adding two numbers together, for example, the ALU will expect to find one of them in the accumulator and this register will also be over-written by the resulting sum.

Active high/low. A statement referring to the logic level of a signal that is required to activate a digital device.

Address. A character or group of characters that identify a register, a particular location in memory or some other data source or destination.

Addressing mode. The component of the instruction op code that specifies how the operand may be located.

ALU. See Arithmetic Logic Unit.

Architecture. The functional capabilities provided by the manufacturer in the design of the device; including such specification as the number and type of registers, internal and external control lines etc.

Arithmetic logic unit (ALU). The digital logic within the processor which performs all arithmetic and logical functions called for by the program.

Assembly language. Code written using the processor's instruction set mnemonics, labels and comments in order to produce an understandable program from which object code may be produced.

BCD (Binary Coded Decimal). A type of positional value code in which each decimal digit is binary coded into 4-bit words.

Bi-directional. Generally refers to interface ports or bus lines that can be used to transfer data in either direction, for example, to or from the microprocessor.

BIT. Contracted form of BInary digiT. May be one or zero. Smallest element of data that may be stored, manipulated or retrieved by a digital computer.

Branch. To depart from the normal sequence of executing instructions in a computer; synonymous with jump.

Branching. A method of selecting, on the basis of the computer results, the next operation to be executed while a program is in progress.

Buffer. A device designed to be inserted between devices to match impedances or equipment speeds, to prevent mixed interactions, to supply additional drive capability, or simply to delay the rate of information flow, classified as inverting or non-inverting.

TERMS — microprocessor

Bug. A program defect or error; also refers to any circuit fault caused by improper design or construction.

Bus. One or more conductors used as a path over which information is transmitted.

Byte. An IBM-developed term used to indicate a specific number of bits treated as a single entity; most often considered to consist of eight bits.

Clock. The basic source of synchronizing signals in most electronic equipment, including computers; a specific device or unit designed to time events.

Conditional jump/branch. A specific instruction which, depending basically upon the result of some arithmetic or logical operation or the state of some flag or indicator, will or will not cause a jump to an instruction in another area of the program.

Condition codes. Single flip-flop type elements within a computer which are set or reset to indicate the result of an operation carried out by the ALU or indicate the status of various functional areas within the processor. Also referred to as flags or status bits.

Control bus. Microprocessor bus carrying input and output control signals for synchronization, external condition sensing, memory management, etc.

CS. Abbreviation for chip select.

Data. A general term used to denote any or all facts, numbers, letters, symbols, etc. which can be processed or produced by a computer.

Data bus. A group of bidirectional lines capable of transferring data to and from the MPU, storage and peripheral devices.

Data port. Physical point at which data enters or leaves a device or system.

Fetch. The period of a computer cycle during which the location of the next instruction is determined, the instruction taken from memory and then entered into the instruction register.

Flags. See **condition codes**.

Flowchart. A graphical method employed by programmers to indicate the stepped procedures of a computer operation; a chart containing all the logical steps in a particular computer program; also called a flow diagram.

Hardware. The metallic or 'hard' components of a computer system in contrast to the 'soft' or programming components; the components of hardware may be active, passive, or both.

Hexadecimal notation. A system of number representation using the base of sixteen. Decimal characters are used for

0-9, after which the characters A to F represent decimal 10-15. May be abbreviated to 'hex'.

Input/output devices. Computer hardware capable of entering data into a computer or transferring data from a computer; abbreviated to I/O.

Instruction. Information which, when properly coded and introduced as a unit into a computer, causes the computer to perform one or more of its operations.

Instruction cycle. That sequence of operations or set of machine cycles which effects one complete instruction.

Instruction set. The total group of characters which, when presented to the computer in their binary form, will result in a pre-determined action or series of actions occurring within the processor.

Interface. Device or circuit which allows two or more systems using different protocols of data representation to communicate accurately.

Interrupt. The suspension of normal operations or programming routines of computers; most often designed to handle sudden requests for service or change; the process of causing the microprocessor to discontinue its present operation and branch to an alternative program routine; also the physical pin-connection line input to the processor.

I/O. Input/Output.

Large-Scale Integration (LSI). A term describing the level of complexity of gates on a single semiconductor chip. LSI chips usually exceed a size of 100 × 100mm and may contain more than 100 000 transistors.

Loop. A self-contained series of instructions in which the last instruction can modify and repeat itself until a terminal condition is reached.

Machine code or language. The basic binary code used by computers; it may be written in hexadecimal notation.

Memory. Stores information for future use; accepts and holds binary numbers.

Memory map. A table showing which addresses in the machine's addressable range have been allocated and to which devices.

Memory mapped input/output. Describes the design practice of assigning a particular memory address or group of addresses to an input/output port or device.

Microcomputer. A general term referring to a complete computing system, consisting of hardware and software and whose main processing blocks have been implemented using semiconductor integrated circuits.

Microprocessor. A principal component of the microcomputer. An integrated circuit capable of performing the major processing tasks associated with the central processing unit (CPU) of larger computers.

Mnemonics. The system of letters adopted by a manufacturer to represent the abbreviated form of the instruction in the instruction set of the computer.

MPU. Microprocessor unit.

Non-volatile. A memory type that retains data even if power has been disconnected.

Object code. The basic program in the form of executable machine code in the language of the particular processor.

Operating code (op code). The specific code relating to the instruction set which selects the required processor operation.

PIO. Parallel input/output circuit. A device containing latches, buffers, flip-flops and other logic circuits needed for versatile input/output between a microcomputer and external circuits. Usually a member of the 'family' of devices associated with the microprocessor in use. May be called Peripheral Interface Adaptor (PIA) or Versatile Interface Adaptor (VIA).

Program. A set of instructions arranged in sequence which direct a computer to perform a desired operation or series of operations.

Programming model. Pictorial representation of the internal register set of processor which may be directly affected by programmed instructions or which may directly affect program execution.

RAM. Memory in which any individual location may be accessed directly. In relation to semiconductor memory devices; it has become synonymous with read and write memory.

Read. The process of copying data from an external device to an internal memory location with respect to the processor.

Register. A flip-flop or group of flip-flops capable of containing one or more data bits or words.

ROM. Read-only memory; programmed by a mask pattern as part of the final manufacturing stage. Information is stored permanently or semi-permanently.

Routine. A set of computer instructions arranged in a correct sequence and used to direct a computer in performing one or more desired operations.

Scratchpad. A 'nickname' for an area of memory used by the program. It is memory containing sub-totals, for example, or various unknowns that are needed for final results.

TERMS — microprocessor

Signed binary numbers. A binary representation of a real number in which the most significant bit is reserved for the sign of the number,
  (MSB = '1' = negative:
  MSB = 0 = positive).

Software. Programs, languages and procedures of a computer system.

Stack. A block of successive memory locations that is accessible from one end on a last-in-first-out (LIFO) basis.

Stack pointer. A register within the processor which contains the address of the next available unused location in the stack. It is automatically adjusted each time a value is added to, or removed from the stack.

Subroutine. A self-contained program that may be 'called' from the main program to the subroutine and then back to main program at the instruction immediately following the subroutine call.

System clock. See clock.

Truth table. Mathematical table showing the Boolean algebraic relationships of variables.

Vector. A software routine's entry address; also the address that points to the beginning of a service routine as it applies to interrupting devices.

Volatile. Storage medium in which data cannot be retained without continuous power dissipation.

Word. A group of binary digits which together represent an element of information, e.g. a number, an ASCII character, etc.

Write. The process of copying data from an internal memory location to an external device with respect to the processor.

APPENDIX – component colour code system

Axial lead inductors

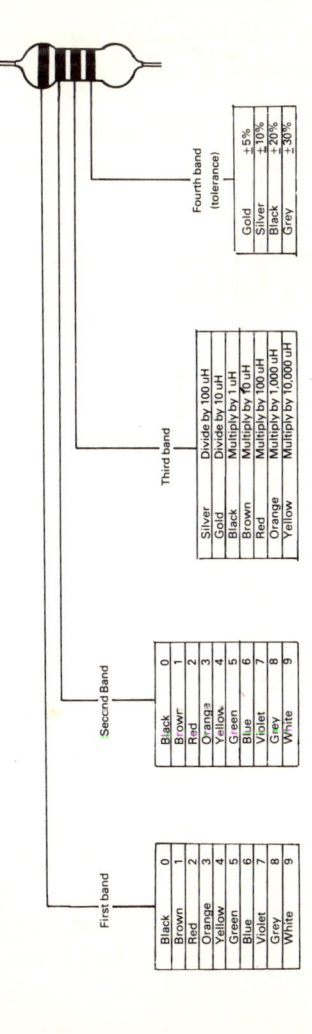

First band

Black	0
Brown	1
Red	2
Orange	3
Yellow	4
Green	5
Blue	6
Violet	7
Grey	8
White	9

Second Band

Black	0
Brown	1
Red	2
Orange	3
Yellow	4
Green	5
Blue	6
Violet	7
Grey	8
White	9

Third band

Silver	Divide by 100 uH
Gold	Divide by 10 uH
Black	Multiply by 1 uH
Brown	Multiply by 10 uH
Red	Multiply by 100 uH
Orange	Multiply by 1,000 uH
Yellow	Multiply by 10,000 uH

Fourth band (tolerance)

Gold	±5%
Silver	±10%
Black	±20%
Grey	±30%

Example: Brown Grey Black Silver = 18uH 10%

APPENDIX — component colour code system

Ceramic plate capacitors

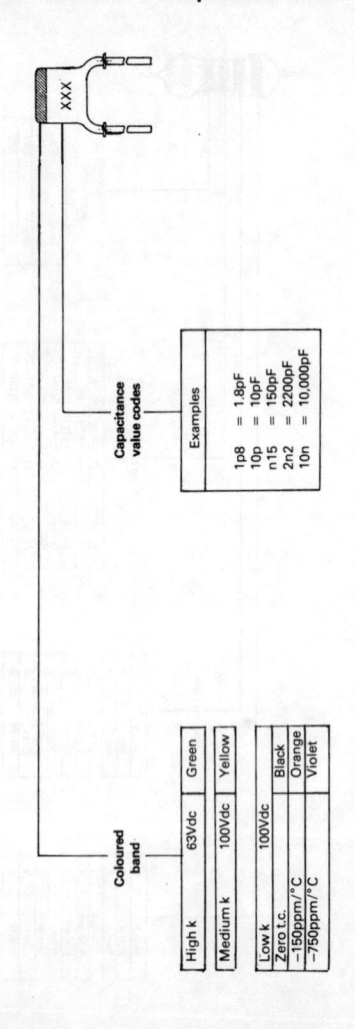

Coloured band		
High k	63Vdc	Green
Medium k	100Vdc	Yellow

Low k	100Vdc	Black
Zero t.c.		Orange
−150ppm/°C		
−750ppm/°C		Violet

Capacitance value codes

Examples

1p8	=	1.8pF
10p	=	10pF
n15	=	150pF
2n2	=	2200pF
10n	=	10,000pF

APPENDIX — component colour code system

Axial lead tubular ceramic capacitors

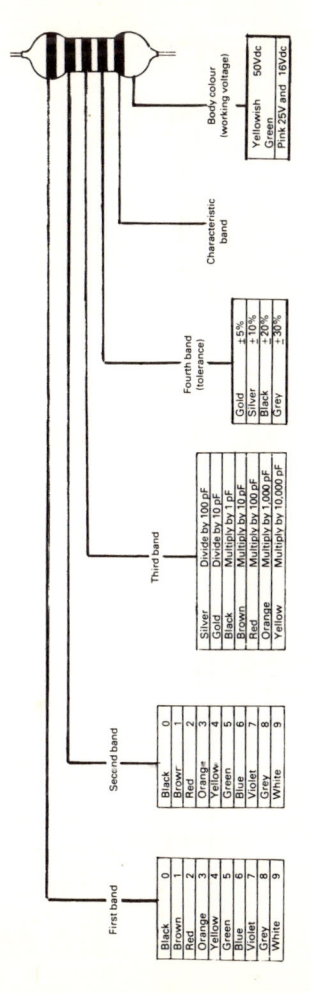

First band

Black	0
Brown	1
Red	2
Orange	3
Yellow	4
Green	5
Blue	6
Violet	7
Grey	8
White	9

Second band

Black	0
Brown	1
Red	2
Orange	3
Yellow	4
Green	5
Blue	6
Violet	7
Grey	8
White	9

Third band

Silver	Divide by 100 pF
Gold	Divide by 10 pF
Black	Multiply by 1 pF
Brown	Multiply by 10 pF
Red	Multiply by 100 pF
Orange	Multiply by 1,000 pF
Yellow	Multiply by 10,000 pF

Fourth band (tolerance)

Gold	±5%
Silver	±10%
Black	±20%
Grey	±30%

Characteristic band

Body colour (working voltage)

Yellowish	50Vdc
Green	
Pink	25V and 16Vdc

Example: Green Blue Gold Silver Black = 5.6pF 10%

APPENDIX — component colour code system

Polyester capacitors

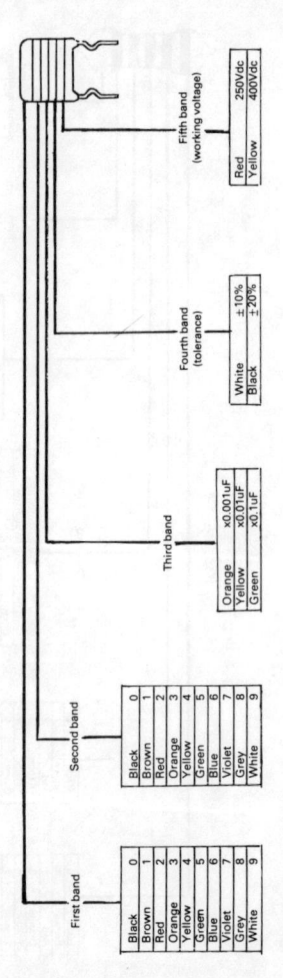

First band

Black	0
Brown	1
Red	2
Orange	3
Yellow	4
Green	5
Blue	6
Violet	7
Grey	8
White	9

Second band

Black	0
Brown	1
Red	2
Orange	3
Yellow	4
Green	5
Blue	6
Violet	7
Grey	8
White	9

Third band

Orange	x0.001uF
Yellow	x0.01uF
Green	x0.1uF

Fourth band (tolerance)

| White | ±10% |
| Black | ±20% |

Fifth band (working voltage)

| Red | 250Vdc |
| Yellow | 400Vdc |

Example: Blue Grey Orange White Red = 0.068uF 10% 250V

APPENDIX – component colour code system

Four-band resistors

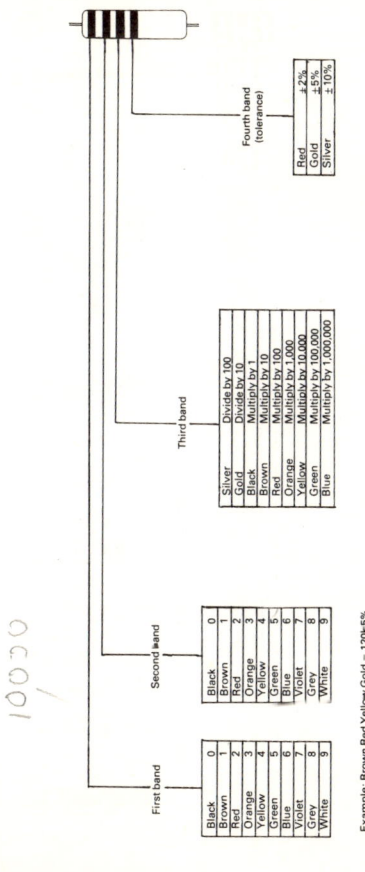

First band

Black	0
Brown	1
Red	2
Orange	3
Yellow	4
Green	5
Blue	6
Violet	7
Grey	8
White	9

Second band

Black	0
Brown	1
Red	2
Orange	3
Yellow	4
Green	5
Blue	6
Violet	7
Grey	8
White	9

Third band

Silver	Divide by 100
Gold	Divide by 10
Black	Multiply by 1
Brown	Multiply by 10
Red	Multiply by 100
Orange	Multiply by 1,000
Yellow	Multiply by 10,000
Green	Multiply by 100,000
Blue	Multiply by 1,000,000

Fourth band (tolerance)

Red	±2%
Gold	±5%
Silver	±10%

Example: Brown Red Yellow Gold = 120k±5%

APPENDIX — component colour code system

Five-band resistors

First band		Second band		Third Band		Fourth band		Fifth band (tolerance)	
Black	0	Black	0	Black	0	Silver	Divide by 100	Brown	±1%
Brown	1	Brown	1	Brown	1	Gold	Divide by 10	Red	±2%
Red	2	Red	2	Red	2	Black	Multiply by 1	Gold	±5%
Orange	3	Orange	3	Orange	3	Brown	Multiply by 10	Silver	±10%
Yellow	4	Yellow	4	Yellow	4	Red	Multiply by 100		
Green	5	Green	5	Green	5	Orange	Multiply by 1,000		
Blue	6	Blue	6	Blue	6	Yellow	Multiply by 10,000		
Violet	7	Violet	7	Violet	7	Green	Multiply by 100,000		
Grey	8	Grey	8	Grey	8	Blue	Multiply by 1,000,000		
White	9	White	9	White	9				

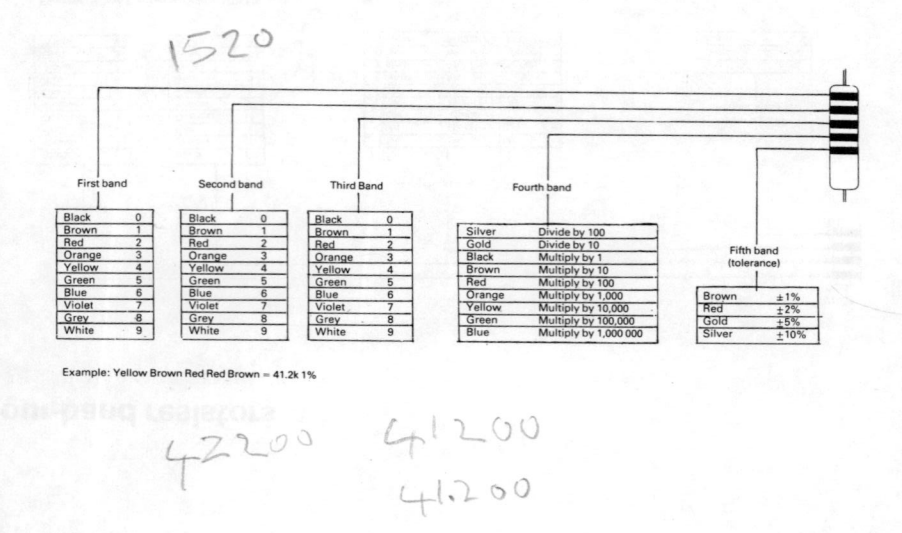

Example: Yellow Brown Red Red Brown = 41.2k 1%

handwritten notes: 152Ω, 4200, 41200, 41200, 41200

APPENDIX – component colour code system 73

Six-band resistors

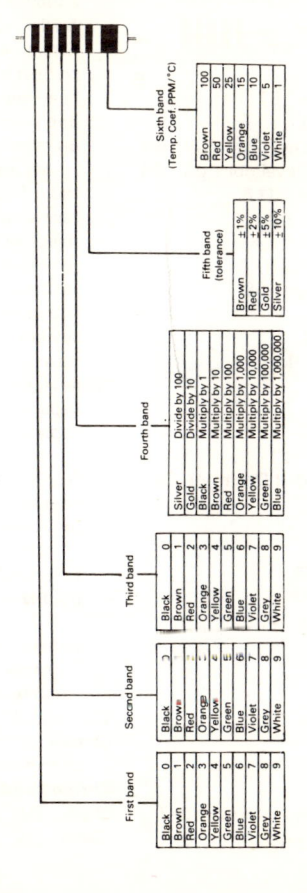

First band		Second band		Third band		Fourth band		Fifth band (tolerance)		Sixth band (Temp. Coef. PPM/°C)	
Black	0	Black	0	Black	0	Silver	Divide by 100	Brown	±1%	Brown	100
Brown	1	Brown	1	Brown	1	Gold	Divide by 10	Red	±2%	Red	50
Red	2	Red	2	Red	2	Black	Multiply by 1	Gold	±5%	Yellow	25
Orange	3	Orange	3	Orange	3	Brown	Multiply by 10	Silver	±10%	Orange	15
Yellow	4	Yellow	4	Yellow	4	Red	Multiply by 100			Blue	10
Green	5	Green	5	Green	5	Orange	Multiply by 1,000			Violet	5
Blue	6	Blue	6	Blue	6	Yellow	Multiply by 10,000			White	1
Violet	7	Violet	7	Violet	7	Green	Multiply by 100,000				
Grey	8	Grey	8	Grey	8	Blue	Multiply by 1,000,000				
White	9	White	9	White	9						

Example: Orange Violet Yellow Brown Brown Red = 3k 74 1% 50PPM

INDEX

Abbreviations 2–8
ACC 48
Accumulator 57
Active high/low 57
Address 57
Addressing mode 57
Aerial symbol 31
AGC (automatic gain control) 48
ALC (automatic level control) 48
Ammeter symbol 31
Antistatic brush 48
Architecture 57
Arithmetic logic unit (ALU) 57
Assembly language 58
Audio head 48
Audio track 48
Auxiliary input 48
Azimuth 48

Back tension 48–9
Battery symbols 21
BCD (Binary Coded Decimal) 58
Belt 49
Betamax 49
Bias signal 49
Bi-directional 58
BIT 58
Brake 49
Branch 58
Branching 58
Buffer 58
Bug 59
Bus 59
Byte 59

Cable distribution systems 40–7
   amplifiers 41–2
   attenuator 47
   equalizers 43–4
   feeder routes 40
   filters 46
   frequency converter 46
   head ends 40
   line power 44–5
   on-selective combiner or separator 47
   radio frequency generators 45
   receiving aerial 40
   selective combiner or separator 46–7
   splitters 42
   subscribers tap-off 43
   system outlet 43
   termination 44
Capacitor symbols 12–13
   capacitor 12
   capacitor with preset adjustment 13
   feed-through capacitor 12
   lead-through capacitor 12
   non-polarized electrolytic capacitor 13
   polarized capacitor 12
   polarized electrolytic capacitor 12
   temperature-dependent capacitor 13
   variable capacitor 13
Capstan 49
Chassis symbols 31
   chassis at offset poential 31
Chroma under 49
Clock 59
Clutch 49

INDEX

Coils and transformer symbols 14–15
   inductor with preset adjustments 15
   inductor with variable inductance 15
   saturable inductor 15
   transformer 14
   winding (inductor, etc.) 14
   winding with core 14
   winding with tappings 14
Compatibility 49
Component colour code system 67–73
   axial lead inductors 67
   axial lead tubular ceramic capacitors 69
   ceramic plate capacitors 68
   five-band resistors 72
   four-band resistors 71
   six-band resistors 73
Condition codes 59
Conditional jump/branch 59
Control bus 59
Control head 50
Control pulses 50
Control track 50
Counter 50
Crosstalk 60
CRT symbols 10–11
   character display tube 10
   coils for electromagnetic deflection 11
   control grid 10
   electron gun 11
   focusing electrode 11
   indirectly heated cathode with associated heater 10
   internal conductive coating 10

CS (chip select) 60

Dark clip 50
Data 60
   bus 60
   port 60
Data processing flowchart symbols 39
   any process function 39
   connector to or from another part of flowchart 39
   connector to or from another part of flowchart on another page 39
   decision operation 39
   input/output function 39
   storage media 39
   terminator, showing entry to and exit from a procedure 39
De-emphasis 50
Delay equaliser 50
Delay line modulator 50
Drop out 51
   compensator 51
Drum 51
Dubbing 51

Earphone symbols 34
Earth symbol 31
E-E 51
End sensors 51
Equalisation 51
Erase 51
   head 51

Fast forward 52
   idler 52
Fetch 60
Flags *see* Condition codes

INDEX

Flowchart 60
Flywheel 52
Freeze frame 52
Fuse symbols 16
   circuit protector 16
   fuse 16

Guard band 52
Guide 52

H 52
Hardware 60
Head-to-tape speed 52
Helical scan 53
Helical wrap 53
   error 53
Hexadecimal notation 60-1
High energy tape 53

Impedance roller 53
Inertia 53
Input/output devices 61
Instruction 61
   cycle 61
   set 61
Interface 61
Interrupt 61
I/O (input/output) 62

Large-Scale Integration (LSI) 62
Logic symbols 35-8
   AND gate 36
   band pass filter 35
   band stop 35
   buffer gate 36
   clocked RS flip-flop 38
   D type flip-flop 38
   edge-triggered 'D' type flip-flop 38
   exclusive 'OR' gate 37
   filter 35
   high pass filter 35
   JK flip-flops 37
   low pass filter 35
   NAND gate 37
   NOR gate 37
   NOT (invertor) gate 36
   OR gate 36
   RS flip-flop 38
   toggle flip-flop 38
Loop 62
Loudspeaker symbols 34

Machine code or language 62
Memory 62
   map 62
   mapped input/output 62
Microcomputer 62
Microphone symbol 33
Microprocessor 63
   terms 57-66
   unit (MPU) 63
Mnemonics 63
Motor discriminator amplifier 53
Motor symbol 32

Non-volatile 63

Object code 63
Operating code (op code) 63

Pause 53-4
   solenoid 54
Peripheral Interface Adaptor (PIA) *see* PIO
Phase shift colour recording 54
PIO (parallel input/output circuit) 63

INDEX

Plug symbols 31
    female 31
    male 31
Pre-emphasis 54
Program 64
Programming model 64

Q 54

RAM 64
Read 64
Recording (writing) symbol 34
Register 64
Relay symbols 33
    relay 33
    relay contact 33
Reproducing (reading, playback) or erasing head symbol 34
Register symbols 16–18
    fixed resistor 16
    fuseable resistor 18
    light sensitive resistor 18
    resistor with inherent non-linear variability 17
    resistor with moving contact 17
    resistor with preset adjustment 16
    resistor with pronounced negative temperature coefficient 18
    resistor with pronounced positive temperature coefficient 18
    variable resistor 16
    voltage dependent resistor 18
    voltage divider with moving contact 17
    voltage divider with preset adjustment 17

Rewind 54
ROM (read-only memory) 64
Rotary transformer 54
Routine 64

Scratchpad 64
Semiconductor symbols 22–30
    avalanche transistor NPN 26
    backward diode 23
    bi-directional diode 23
    bi-directional diode thyristor 24
    bi-directional triode thyristor 25
    clipper diode 23
    diode 22
    diode used as capacitance device 22
    diode where use is made of temperature dependence characteristic 22
    IG Fet Depletion type. N-channel Substrate internally connected 28
    IG Fet Depletion type. P-channel 28
    channel Substrate Internally connected 28
    IG Fet Depletion type. P-channel 28
    IG Fet Depletion type. P-channel Substrate connection brought out 28
    IG Fet Depletion type. Single gate N-channel 28
    IG Fet Depletion type two gate N-channel 29
    IG Fet Enhancement type single-gate P-channel 29

Semiconductor symbols (cont'd)
- IG Fet Enhancement type two-gate N-channel 29
- JUG Fet N-channel 27
- JUG Fet P-channel 28
- light-emitting diode 30
- light-sensitive diode 30
- photo voltage cell 30
- PNP photo device 29
- reverse-blocking diode thyristor 23
- reverse-blocking triode thyristor N-gate (anode controlled) 24
- reverse-blocking triode thyristor P-gate (cathode controlled) 24
- reverse-blocking thyristor tetrode (SCS) 25
- reverse-conducting diode thyristor 24
- reverse-conducting triode thyristor N-gate (anode controlled) 24
- reverse-conducting triode thyristor P-gate (anode controlled) 25
- thyristor 23
- transistor NPN 26: with collector connected to envelope 26
- transistor PNP 26
- transistor with ohmic connection to the I region PNIN 27
- transistor with ohmic connection to the I region PNIP 27
- transistor with transverse biased base NPN 27
- tunnel diode 22
- turn-off triode thyristor N-gate (anode controlled) 25
- turn-off triode thyristor P-gate (cathode controlled) 25
- unijunction transistor N-type base 27
- unijunction transistor P-type base 26
- varicap diode 22
- zener diode 22

Servo 54
Signed binary numbers 65
Skew error 55
Slant azimuth 55
Software 65
Stack 65
- pointer 65
Stop solenoid 55
Subroutine 65
Switch symbols 32
- electronically operated switch 33
- normally closed switch 32
- normally open switch 32
Symbols
- Greek letters 9
- mathematical 9
System clock *see* Clock

Tape wrap *see* Helical wrap
Tension *see* Back tension
Timebase error 55
Tracking 55
Transformers *see* Coils and transformer symbols
Truth table 65

Vector 65

INDEX

Versatile Interface Adaptor (VIA) *see* PIO
VHS 55
Video heads 55
Video recording terms 48–56
Volatile 66
Voltmeter symbol 32

Waveform symbols 19–20
   negative going 19
   pulse alternating current 19
   pulse amplitude modulation 20
   pulse code modulation 20
   pulse duration modulation 20
   pulse frequency modulation 20
   pulse interval modulation 20
   pulse position or pulse phase modulation 20
   pulse positive going 19
   sawtooth 19
   step negative going 19
   step positive going 19
White clip 56
Word 66
Wow and flutter 56
Write 66
Writing speed 56